HITLER'S WAR MACHINE

PANZER IV
THE WORKHORSE OF THE PANZERWAFFE

EDITED AND INTRODUCED BY
BOB CARRUTHERS

Pen & Sword
MILITARY

This edition published in 2013 by
Pen & Sword Military
An imprint of
Pen & Sword Books Ltd
47 Church Street
Barnsley
South Yorkshire
S70 2AS

First published in Great Britain in 2012 in digital format by
Coda Books Ltd.

Copyright © Coda Books Ltd, 2012
Published under licence by Pen & Sword Books Ltd.

ISBN 978 1 78159 205 2

A CIP catalogue record for this book is
available from the British Library

All rights reserved. No part of this book may be reproduced or transmitted in any form or by any means, electronic or mechanical including photocopying, recording or by any information storage and retrieval system, without permission from the Publisher in writing.

Printed and bound by
CPI Group (UK) Ltd, Croydon, CR0 4YY

Pen & Sword Books Ltd incorporates the Imprints of Pen & Sword Aviation, Pen & Sword Family History, Pen & Sword Maritime, Pen & Sword Military, Pen & Sword Discovery, Pen & Sword Politics, Pen & Sword Atlas, Pen & Sword Archaeology, Wharncliffe Local History, Wharncliffe True Crime, Wharncliffe Transport, Pen & Sword Select, Pen & Sword Military Classics, Leo Cooper, The Praetorian Press, Claymore Press, Remember When, Seaforth Publishing and Frontline Publishing

For a complete list of Pen & Sword titles please contact
PEN & SWORD BOOKS LIMITED
47 Church Street, Barnsley, South Yorkshire, S70 2AS, England
E-mail: enquiries@pen-and-sword.co.uk
Website: www.pen-and-sword.co.uk

CONTENTS

INTRODUCTION 4

CHAPTER 1: THE WORKHORSE OF THE PANZERWAFFE .. 8

CHAPTER 2: COMBAT 12

   THE CONTEMPORARY VIEW #1 12

   THE CONTEMPORARY VIEW #2 19

   THE CONTEMPORARY VIEW #3 26

   THE CONTEMPORARY VIEW #4 32

   THE CONTEMPORARY VIEW #5 41

CHAPTER 3: DEVELOPMENT HISTORY 44

   THE CONTEMPORARY VIEW #6 54

   THE CONTEMPORARY VIEW #7 57

   THE CONTEMPORARY VIEW #8 63

   THE CONTEMPORARY VIEW #9 64

   THE CONTEMPORARY VIEW #10 69

MORE FROM THE SAME SERIES 90

INTRODUCTION

This book forms part of the series entitled 'Hitler's War Machine.' The aim is to provide the reader with a varied range of materials drawn from original writings covering the strategic, operational and tactical aspects of the weapons and battles of Hitler's war. The concept behind the series is to provide the well-read and knowledgeable reader with an interesting compilation of related primary sources combined with the best of what is in the public domain to build a picture of a particular aspect of that titanic struggle.

I am pleased to report that the series has been well received and it is a pleasure to be able to bring original primary sources to the attention of an interested readership. I particularly enjoy discovering new primary sources, and I am pleased to be able to present them unadorned and unvarnished to a sophisticated audience. The primary sources such as 'Die Wehrmacht' and 'Signal', speak for themselves and the readership I strive to serve is the increasingly well informed community of reader/historians which needs no editorial lead and can draw its own conclusions. I am well aware that our community is constantly striving to discover new nuggets of information, and I trust that with this volume I have managed to stimulate fresh enthusiasm and that at least some of these facts and articles will be new to you and will provoke readers to research further down these lines of investigation, and perhaps cause established views to be challenged once more. I am aware at all times in compiling these materials that our relentless pursuit of more and better historical information is at the core our common passion. I trust that this selection will contribute to that search and will help all of us to better comprehend and understand the bewildering events of the last century.

In order to produce an interesting compilation giving a flavour

The Panzer IV from the business end. The practice of adding the names of sweethearts to the vehicle was widespread in this unit.

of events at the tactical and operational level I have returned once more to the US Intelligence series of pamphlets, which contain an intriguing series of contemporary articles on weapons and tactics. I find this series of pamphlets particularly fascinating as they are written in the present tense and, as such, provide us with a sense of what was happening at the face of battle as events unfolded.

The Panzer IV was originally designed as an infantry support tank with a unique tactical role. The Panzer IV was not designed to take part in tank vs tank combat. Although the Panzer IV initially had relatively thin armour, it carried a powerful 75mm gun and could match any other tank at that time. The prototype of the Panzer IV was given the code name Bataillonfuhrerwagen. The Panzer IV was ordered by Hitler from Krupp, MAN and Rheinmetall Borsig to weigh in at 18 tonnes with a top speed of 35 km/hr. The Krupp design - the VK 200 1 (K) - was eventually selected to enter into full-scale production in 1935. Along with the Panther, it was to become the main combat tank of the Third Reich.

The PzKpfw IV was perceived as the 'workhorse' of all the Panzer

Panzer IV tanks of the SS-Division 'Hitlerjugend' on parade February 1944.

divisions and more were produced than any other variant in the 1933-1945 period. The Ausf. A was built as a pre-production vehicle and only 35 were produced. The modifications from this gave rise to the Ausf. B which emerged in 1938 with an increased frontal armour thickness and a six-speed gearbox, which enhanced its cross-country performance. That same year Krupp-Gruson produced the Ausf. C and 134 of this model were in production until 1939.

The Ausf. D/E saw an upgrading of its armour thickness and improved vision blocks for the driver. The Ausf. E was the first of the Panzer IV fitted with turret mounted stowage bins. The Ausf. F(1), produced between 1941-1942 was the last Panzer IV to be based on the short version chassis. 25 of the F Is were converted into Ausf F2s (it had the British nickname of "Mark IV Special" because, with its high velocity 75mm main armament it was far superior to any other tank at the time). It was followed by the modified version of the Ausf. G in May 1942.

The Ausf. H, introduced in April 1943, was exclusively armed with a newer version of the 75mm KwK 40 L/48 gun and was fitted with steel/wire armour skirts. Over 3,770 of the P/zKpfw IV Ausf H were made and saw action.

As late as 1945 the last model, the Ausf J, was an effective weapon in the hands of an experienced crew. A selected number of the Ausf H and J were also converted into command tanks or observation tanks towards the end of the war period.

The Panzer IV was the only German tank to stay in production throughout the war. It was the real workhorse of the German army and was deployed on every front. Due to its efficient armament, robust armour and outstanding reliability, it was preferred by crews over the Panther, Tiger and King Tiger. The Panzer IV was the most widely exported tank in German service, with around 300 sold to partners such as Finland, Romania, Spain and Bulgaria. After the war, the French and Spanish sold dozens of Panzer IVs to Syria, where they saw combat in the 1967 Six-Day War.

Thank you for buying this volume in the series we hope it will spur you on to try the others.

Bob Carruthers
Edinburgh 2012

A Panzer IV Ausf. E showing signs of multiple hits to the turret, including the gun barrel.

A Panzer IV Ausf.A undergoing testing during 1938.

CHAPTER 1
THE WORKHORSE OF THE PANZERWAFFE

The Panzerkampfwagen IV (Pz.Kpfw. IV) Sd Kfz 161, commonly known as the Panzer IV, was a medium tank developed in Nazi Germany in the late 1930s and used extensively during the Second World War. Its ordnance inventory designation was Sd.Kfz. 161.

Designed as an infantry-support tank, the Panzer IV was not originally intended to engage enemy armor as this function was intended to be performed by the lighter Panzer III. However, by 1941, the flaws of pre-war doctrine had become apparent and in the face of the Soviet T-34 tanks, the Panzer IV soon assumed the tank-

PzKpfw IV Ausf. D

Panzer IV Ausf.H, Russia 1944

A field conference with a column of factory fresh Panzer IV's in the background, Russia 1944.

fighting role instead of the obsolete Panzer III which was too small to cope with a high velocity main armament. The Panzer IV chasis was robust and strong enough to accept a number of upgrades in armour and armament. As a result it was destined to become most widely manufactured and deployed German tank of the Second World War. The Panzer IV was used as the base for many other fighting vehicles, including the Sturmgeschütz IV tank destroyer, the Wirbelwind self-propelled anti-aircraft weapon, and the Brummbär self-propelled gun, amongst others.

Robust and reliable, it saw service in all combat theaters involving German forces, and has the distinction of being the only German tank to remain in continuous production throughout the war, with over 8,800 produced between 1936 and 1945. Upgrades and design modifications, often made in response to the appearance of new Allied tanks, extended its service life. Generally these involved increasing the Panzer IV's armour protection or upgrading its weapons, although during the last months of the war with Germany's pressing need for rapid replacement of losses, design changes also included retrograde measures to simplify and speed manufacture.

CHAPTER 2
COMBAT

The following article is taken from the US wartime publication Intelligence Report. It provides a clear account of the duties of the crew from a widely used publication.

THE CONTEMPORARY VIEW NO. I

CREW AND COMMUNICATIONS OF GERMAN MARK IV TANK

Extracted from
Technical and Tactical Trends
No. 12, November 19th, 1942

The duties of the various crew members of the Mark IV tank are generally similar to those performed by the crews of our own medium M3 and M4 tanks. A German training pamphlet captured in Libya gives the following details on the crew duties and communications of the Mark IV.

A. DUTIES OF THE CREW

The crew consists of five men: a commander, gunner, loader, driver, and radio operator. The latter is also the hull machine-gunner.

(1) Tank Commander

The tank commander is an officer or senior NCO and

The Commander was the most important component of the crew. He was the eyes and ears and the decision maker.

is responsible for the vehicle and the crew. He indicates targets to the gunner, gives fire orders, and observes the effect. He keeps a constant watch for the enemy, observes the zone for which he is responsible, and watches for any orders from the commander's vehicle. In action, he gives his orders by intercommunication telephone to the driver and radio operator, and by speaking tube and touch signals to the gunner and loader. He receives orders by radio or flag, and reports to his commander by radio, signal pistol, or flag.

(2) Gunner

The gunner is the assistant tank commander. He fires the turret gun, the turret machine gun, or the submachine gun as ordered by the tank commander. He assists the tank commander in observation.

(3) Loader

This crew member loads and maintains the turret

Panzer crew member and Panzer IV Ausf. B

armament under the orders of the gunner. He is also responsible for care of ammunition, and when the cupola is closed, gives any necessary flag signals. He replaces the radio operator if the latter becomes a casualty.

(4) Driver

The driver operates the vehicle under the orders of the tank commander or in accordance with orders received by radio from the commander's vehicle. So far as possible he assists in observation, reporting through the intercommunication telephone the presence of the enemy or of any obstacles in the path of the tank. He watches the gasoline consumption and is responsible to the tank commander for the care and maintenance of the vehicle.

(5) Radio Operator

He operates the radio under the orders of the tank commander. In action, and when not actually transmitting, he always keeps the radio set to "receive." He operates the intercommunication telephone and takes down any useful messages he may intercept. He fires the machine gun mounted in the front superstructure. If the loader becomes a casualty, the radio operator takes over his duties.

B. COMMUNICATIONS

The following means of communication may be used:

(1) External: radio, flag, hand signals, signal pistol, and flashlight.

(2) Internal: intercommunication telephone, speaking tube, and touch signals.

For the radio, the voice range between two moving vehicles is about 3 3/4 miles and CW about 6 1/4 miles.

The flag is used for short-range communications only, and the signal pistol for prearranged signals, chiefly to other arms.

The radio set, in conjunction with the intercommunication telephone, provides the tank commander, radio operator, and driver with a means for external and internal voice communication, the same throat microphones and telephone receiver headsets being used for both radio and telephone.

When the control switch on the radio is set at EMPFANG (receive) and that on the junction box of the intercommunication telephone at BORD UND FUNK (internal and radio), the commander, radio operator, and driver hear all incoming radio signals. Any one of them can also speak to the other two, after switching his microphone into circuit by means of the switch on his chest.

For radio transmission, the switch on the set is adjusted to TELEPHONIE. The telephone switch may be left at BORD UND FUNK. Either the tank commander or the radio operator can then transmit, and they and the driver will all hear the messages transmitted. Internal communication is also possible at the same time, but such conversation will also be transmitted by the radio.

If the radio set is disconnected or out of order, the telephone switch may be adjusted to BORD (internal). The tank commander and driver can then speak to one another, and the radio operator can speak to them, but cannot hear what they say. The same applies when a radio receiver is available but no transmitter, with the difference that incoming radio signals can then be heard by the radio operator.

The signal flags are normally carried in holders on the left of the driver's seat. When the cupola is open, flag signals are given by the tank commander, and when it is closed, the loader raises the circular flap in the left of the turret roof and signals with the appropriate flag through the port thus opened.

The signal pistol is fired either through the signal port in the turret roof, through the cupola, or through one of the vision openings in the turret wall. The signal pistol must not be cocked until the barrel is already projecting outside the tank. It is only used normally when the vehicle is stationary. Its main use is giving prearranged signals to the infantry or other troops.

When traveling by night with lights dimmed or switched off altogether, driving signals are given with the aid of a dimmed flashlight. The same method is also employed when tanks are in a position of readiness and when in bivouac.

Orders are transmitted from the tank commander to the gunner by speaking tube and by touch signals. The latter are also used for messages from the commander to the loader, and between the gunner and loader.

A British Crusader passes an abandoned Panzer IV tank, Libyan desert 1941

The Panzer IV was vulnerable to close assault particularly in the air intakes and the Allies were quick to recognise this-

'When enemy armoured force vehicles are attacked at close quarters with incendiary grenades, the air louvres are very vulnerable. It is therefore important that differentiation be made between "inlet" and "outlet" ducts, since obviously a grenade thrown against an exhaust opening will be less effective than one aimed at an inlet, which will draw the inflammable liquid into the vehicle. If the engine is not running, all openings are equally vulnerable.

'In general, it may be said that in the Pz Kw II and III tanks the best targets are the flat top-plates of the rear superstructures, since the air intakes are located there. The side louvres in these tanks are invariably protected by a vertical baffle. On the Pz Kw IV, the left side ports are intake and thus more vulnerable than the right-hand exhaust ports.'

Faced with these and other threats on the battlefield the German designers were quick to improve the armour on the Panzer IV, but by 1943 the Allies were aware of these developments.

THE CONTEMPORARY VIEW NO.2
INCREASED PROTECTION ON PzKw 3 AND 4

Extracted from Technical and Tactical Trends No. 25, May 20th, 1943

The history of the changes in the light medium PzKw 3 and 4 demonstrates how fortunate the Germans were in having a basic tank design that could be improved as battle experience indicated, for a basic design can be improved and still remain familiar to the users. Furthermore, the problems of maintenance and supply of parts are greatly reduced--and these problems are a major factor in keeping tanks ready for operational use.

PZKW 4

(1) Early Models

The PzKw 4, a slightly heavier tank than the 3, has passed through much the same line of development. Little is known about the models A, B, and C of this tank, but Model D was in use during the greater part of the period 1940-43. Specimens of armour cut from Model D have been examined. Of these, only the front plate of the hull appears to be face-hardened; this plate is carburized. All of the plates were high-quality, chromium-molybdenum steel, apparently made by the electric-furnace process.

The first increase in the armour of this tank was

A Panzer IV Ausf.C still in service with the GrossDeutchsland Division in Novemeber 1943.

reported in 1941, when it was observed that additional plates had been bolted over the basic front and side armour. The additional plates on the front were 1.18 inches thick, making a total of 2.36 inches, and those on the sides were .79 inches thick, making a total of 1.57 inches. In its early stages, this addition was probably only an improvised measure for increasing the armour protection of existing PzKw 4 models in which the thickest armour was only 1.18 inches.

(2) Model E

In Model E, which had 1.96 inches of single-thickness nose plate, the fitting of additional armor on the front of the superstructure and on the sides of the fighting compartment was continued. Although the arrangement of the additional side armor on this model appears to have been standardized, that on the front superstructure was by no means uniform.

A column of Panzer IV pass a transport-column, Russia 1944.

Panzer IV Auf.D in France 1940.

A Panzer IV with supporting infantry advancing towards the Nettuno bridgehead during 1944.

Three PzKw 4 tanks have recently been examined. In each case, extra armor had been fitted to the vertical front plate carrying the hull machine gun and driver's visor. It had also been added to the sides of the fighting compartment both above and below the track level. The extra protection above the track level extended from the front vertical plate to the end of the engine-compartment bulkhead. It was thus 110 inches long and 15 inches deep. The pieces below the track level were shaped in such a way as to clear the suspension brackets. They were 90 inches long and 30 inches deep. All this extra side protection was .97 inch in thickness.

The vertical front plate was reinforced in three different ways. On one tank, two plates were used; one over the plate carrying the hull machine gun, this additional plate being cut away to suit the gun mounting, and the other plate over the driver's front plate, cut to shape to clear his visor. On the second tank, the arrangement around the hull gun was the same, but the extra protection around the driver's visor consisted of two rectangular plates, one on each side of the visor, there being no extra plate immediately above the visor. On the third tank, the only additional front armor was the plate around the hull machine gun. No additions had been made to the driver's front plate. In all cases, the extra frontal plating was 1.18 inches thick; the nose plate was unreinforced, but it was 1.97 inches thick, and the glacis plate was .97 inch thick. The final drive casings of PzKw 4 tanks of this period were also sometimes reinforced by .79-inch protecting rings. The additional plates on the front were face-hardened.

It is probable that the reinforced armor on the front superstructure of this model will compare closely with that on the corresponding parts of the PzKw 3 of 1941 and that the 1.96-inch nose plates will not differ substantially from those on the more recent PzKw 3's of June 1942, known as "Model J."

The reinforced (.79 inch plus .79 inch) side armor has, however, no counterpart in any PzKw 3 model. The additional plates are of homogeneous quality and have a Brinell hardness of about 370 on the front surface.

(3) Model F

Towards the end of 1941 the Germans introduced a PzKw 4, Model F, having 1.96-inch frontal armor (gun mantlet, front superstructure and hull nose-plates) and 1.18-inch side armor. In this and many other respects, the Model F conforms more closely than its predecessors to the corresponding model of the PzKw 3 (in this case PzKw 3 Model J). So far, the armor of the PzKw 4 Model F has not been examined to ascertain its chemical and ballistic properties, but there is a strong probability that these do not differ greatly from those of the PzKw 3, Model J.

(4) Model G

This model which mounts the long 75-mm gun, Kw.K 40, was first encountered in June 1942. It is reported from the Middle East that its armour is the same as that of Model F; namely 1.96 inches on the front, and 30 mm (1.18 inches) on the sides.

An excellent study of a group of Panzer IV Ausf.E gathered together on exercise in France during 1943. Bringing together this number of vehicles in such close proximity would not have been permitted in Russia.

Women machining tank parts in the Krupp factory.

In addition to the increase in armor it was necessary to up-gun the tank by introducing a high velocity main armament which gave the Panzer IV its tank killing power. Not surprisingly the Allies were soon aware of this development and the intelligence was quickly spread through the regular channels.

> THE CONTEMPORARY VIEW NO.3
> ## NOTES ON THE PzKw 4
> Extracted from
> Technical and Tactical Trends
> No. 27, June 17th, 1943
>
> The PzKw 4 is the German standard medium tank. It weighs about 22 tons. With the exception of the principal armament, the more recent models of this tank embody essentially the same features. The change in armament consists of a long-barreled 75-mm gun, the 7.5-cm Kw K. 40, being fitted in place of the short-barreled 75-mm gun (see Tactical and Technical Trends, No. 20, p. 10).
>
> The following information on the new PzKw 4 is based on a tank captured in North Africa.
>
> a. Suspension and Armor
>
> The tank has eight small bogie wheels, mounted and sprung in pairs by quarter-elliptic springs, a front sprocket, a rear idler, and four return rollers on each side. The track is of steel, as is usual in German tanks.
>
> The armor probably is as follows: front, back, and turret 1.95 in.; sides 1.18 in.; back and top .39 to .79 in. [Later details indicate that the armor arrangement on

The Panzer IV Aus.F, seen here in the Army Group North sector during the summer of 1942, was the last of the short barrelled tanks to see action.

current models of PzKw 4 is the same.] Sand bags were carried on top of the turret for additional protection from air attack. (German tanks often carry sand bags and additional lengths of track as added protection.)

b. Dimensions and Performance

The tank is 19 ft. 6 in. long, 9 ft. 4 in. wide, and 8 ft. 9 in. high, with a ground clearance of 16 inches. It can cross a 9-foot trench, negotiate a 2-foot step, climb a 27-degree gradient, and ford to a depth of 2 ft. 7 in. The theoretical radius of action is 130 miles on roads and 80 miles cross-country.

c. Engine

The tank is powered with a Nordbau Model V-12, four-stroke, gasoline engine, developing 320 hp. It has overhead cams, one for each bank of engines, and magneto ignition. There are two Solex down-draught carburetors, and twin radiators, with a fan for each, mounted on the right-hand side of the engine. An inertia starter is fitted. [An inertia starter is a starter equipped with its own independent fly-wheel to build up starting inertia.] The fuel capacity is 94 gallons for the engine and 20 gallons for the 2-cylinder turret-drive auxiliary engine.

d. Clutch, Brake, and Drive

The clutch is incorporated in a gear-box which is of the ordinary type with 6 forward speeds and reverse. The brakes, operating on epicyclic gears, are air-cooled and hydraulically operated. The drive is through the engine, drive shaft, clutch, gear box, bevel drive, steering system, final reduction drive, and sprockets.

e. Instruments

Instruments include a revolution counter (tachometer) to 3,200 rpm with 2,600 to 3,200 in red, speedometer

A Panzer IV Ausf. B rolls into action during the Polish campaign, September 1939.

Panzer IV on manoeuvres in Greece during 1942.

Pz. Kw. 4

to 50 kph (31 mph), odometer (mileage indicator), a water temperature gauge, and two oil pressure gauges reading to 85 lbs. per sq. in. The tank is fitted to take an electric gyrocompass on the left side of the driver.

f. Armament

The tank mounts the long-barreled 75-mm gun and two model 34 machine guns, one fixed coaxially on the right side of the gun, and the other one set in the hull firing forward. While reports vary, it is thought that the gun will penetrate 2 inches of homogeneous armor at about 2,500 yards at 30 degrees. The breech is of the vertical sliding type. Firing is electric, with a safety device which prevents firing if the breech is not closed, the gun not fully run out, or the buffer not full. The traverse is by hand, or by power from a 2-cylinder, 9-hp auxiliary gasoline engine directly coupled to a generator, which supplies current to the turret traversing motor. The turret floor rotates. Eighty-three rounds of 75-mm AP or HE and smoke are carried. Five smoke candles may be carried on a rack at the rear of the tank. These candles are released from inside by a wire

A column of Panzer IVs at the halt during a road march through Yugoslavia.

cable. Twenty-seven belts of 75 rounds each are carried for the machine guns.

g. Radio Equipment

Intercommunication is by radio-telephone. The aerial may be raised or lowered from inside the tank. The set is situated over the gear box on the left side of the hull gunner. Below the 75-mm gun is situated an insulated aerial guard which deflects the aerial when the turret is traversed.

h. Crew

The crew numbers five: driver, hull-gunner and radio operator, commander, gunner, and loader.

The tactical application of the Panzer IV was also of great interest and the Allied intelligence services were delighted by the capture of a German training manual.

THE CONTEMPORARY VIEW NO.4
COMBAT TACTICS OF GERMAN MEDIUM TANK COMPANIES
Extracted from Technical and Tactical Trends No. 26, June 3rd, 1943

a. General

The following combat instructions for PzKw 4 units have been condensed from a German document. They give an excellent idea of recent enemy tank tactics.

A group of Panzer IVs rolls forward into action during the winter of 1943.

A superb study of the Panzer IV in service with the Hitler Jugend Division. Note the Zimmerit covering.

b. Individual Tactics

(1) In view of the small amount of ammunition carried, the gun is normally fired at the halt in order to avoid waste. The machine guns mounted in turret and hull may be effectively fired up to 800 yards against mass targets, such as columns, reserves, limbered guns, etc.

(2) As soon as each target has been put out of action, or as soon as the attacking German infantry are too near the target for tanks to fire with safety, the tanks move forward by bounds of at least 200 to 300 yards. When changing position, drivers must take care to keep correct position in the tactical formation.

(3) Single tanks may be used for supporting action against prepared positions. The tank will normally move from a flank under cover of smoke. Embrasures will be engaged with AP shell. During action, it will be necessary to blind neighbouring defences by smoke.

Tanks will normally fire at prepared defences from at most 400 yards' range. Assault detachments work their way forward, and once lanes have been cleared through the antitank defences, the tank will follow and engage the next target. Close cooperation between tank and assault detachment commanders is essential. Light and other signals must be prearranged. Single tanks can also be used in fighting in woods and for protection of rest and assembly areas.

c. Platoon Tactics

(1) During the attack, medium platoons move forward in support of the first wave; one half of the platoon gives covering fire while the other half advances. The whole platoon seldom moves as a body.

(2) The platoon commander directs by radio, and he can control fire by radio or by firing guiding-rounds on particular targets.

(3) Antitank weapons will normally be engaged from the halt. If the nearest antitank weapon can be dealt with by the light platoon, the medium platoon will engage more distant antitank weapons or blind them. Artillery will be attacked in the same manner as antitank weapons. Enfilading fire is particularly recommended.

(4) If friendly light tanks encounter enemy tanks in the open, the medium platoon should immediately engage them with smoke-shell in order to allow the lights to disengage and to attack the enemy from a flank.

(5) Moving targets and light weapons should be engaged with machine guns or by crushing; mass targets with HE.

A Panzer IV with supporting infantry and Sturmgeschutz move into action February 1944.

Panzer IV Ausf.H in the snow, Russia 1943.

(6) Against prepared defenses, the procedure is as mentioned in Paragraph b (3). When the whole platoon is employed, the advance can be made by mutual fire and smoke support. When the position is taken, the platoon covers the consolidation by smoke and fire. The platoon only moves forward again after the enemy weapons in the prepared position have been knocked out.

(7) In street fighting a medium platoon may be employed in the second echelon to give support. Nests of resistance in houses may be cleaned up with the help of the tanks' guns, and lightly built houses can be crushed.

(8) If a front-line tank formation is ordered to hold an objective until the arrival of infantry, protection will be given by the medium platoon, which will take up position on high ground with a large field of fire.

d. Company Tactics

(1) When medium platoons are attached to light companies, they work on the latter's radio frequency, and not on that of their own medium company.

(2) Reserve crews follow immediately behind the combat echelon and move back to join the unit trains only after the beginning of an engagement. They come forward again as soon as the battle is over. Reliefs must be so arranged that drivers take over refreshed before each action, that is, on leaving the assembly area.

(3) The repair section, commanded by an NCO, travels with the combat echelon until the beginning of the battle.

Grenadiers crammed aboard a Panzer IV, Russia February 1944.

(4) The company commander moves at the head of his company until the leading platoons have gone into action, when he operates from a temporary command post with unimpeded observation of the battle area. Keeping direction and contact are the responsibility of company headquarters personnel while the commander is at the head of his company.

(5) In the attack, the normal formations are a broad wedge - Breitkeil - [One platoon echeloned to the right, one to the left, and one in line to form the base of the triangle, with apex forward], or line with extended interval (geoffnete Linie). Effective fire of the whole company may be obtained if the rear elements give overhead fire, or if they fill up or extend the front of their company to form line.

(6) For tank-versus-tank actions, the company, where possible, should be employed as a whole. When enemy tanks appear, they must be engaged at once and

other missions dropped. If time allows, the battalion commander will detach the medium platoons that have been attached to light companies and send them back to the medium company. In all situations, medium tanks should endeavor to have the sun behind them.

(7) During the pursuit, the medium company will be employed well forward in order to take full advantage of the longer range of its HE shell.

e. Miscellaneous

(1) The light tank platoon of battalion headquarters company guides the medium company on the march, and when going in to rest or assembly positions. If the medium company is moving on its own, one section of a light tank platoon may be attached to it.

(2) Parts of the antiaircraft platoon of the headquarters company may be allotted to the medium company.

(3) Tank repairmen move directly behind the combat echelons. The recovery platoon is responsible for towing away those tanks which cannot be attended to by the repair section. The recovery platoon is under the orders of the technical officer, who has under his control all equipment and spare-parts trucks of the tank companies, which may follow by separate routes as prescribed by him.

The crew of a Panzer IV Ausf.D on all round aerial observation, Russia spring 1942.

The brand new Panzer IV Ausf.G on exercise in Greece.

Two destroyed Panzer IVs of the 20th Panzer Division photgrphed just outside Bobruisk.

As the war progressed more and more intelligence became available and the methods available to combat the Panzer IV were constantly revised. An intelligence report entitled "Vulnerability of German Tank Armor" was published in Tactical and Technical Trends.

THE CONTEMPORARY VIEW NO.5

VULNERABILITY OF GERMAN TANK ARMOR

Extracted from
Technical and Tactical Trends
No. 8, September 24th, 1942

British forces in the Middle East have recently carried out tests with captured German tanks in order to determine the effectiveness of British and U.S. weapons against them.

The 30-mm front armor of the original German Mark III tank (see this publication No. 3, page 12) is apparently a plate of machinable-quality silico manganese. The additional 30- or 32-mm plates which have been bolted onto the basic 30-mm armor are of the face-hardened type. This total thickness of 60 to 62 mm stops the British 2-pounder (40-mm) AP ammunition at all ranges, breaking it up so that it only dents the inner plate. The U.S. 37-mm projectile, however, with its armor-piercing cap, penetrates at 200 yards at 70°. Against the 6-pounder (57-mm) AP and the 75-mm SAP, this reinforced armor breaks up the projectile down to fairly short ranges, but the armor plate itself cracks and splits fairly easily, and the bolts securing it are ready to give way after one or two hits. If 75-mm capped shot is used, however, such as the U.S. M61 round, the armor can be pierced at 1,000 yards at 70°.

Similar results may be expected against the reinforced armor of the Mark IV.

The new Mark III tank has a single thickness of 50-mm armor on the front, and this was found to be of the face-hardened type. The 2-pounder AP projectile penetrates by shattering the hardened face, but the projectile itself breaks up in the process and the fragments make a hole of about 45 mm. The 37-mm projectile does not shatter during penetration, which is secured at ranges up to 500 yards at 70°. The 50-mm plate is softer than the reinforced 32-mm plates being 530 Brinell on the face and 375 on the back. This plate is not particularly brittle and there is very little flaking.

In tests carried out against the side armor of both the old and new models of Mark III tanks, it was found that this armor showed signs of disking at the back. There is also internal petaling. This, and the condition of the front, which is flaked back at 45° for a short distance, indicates that the heat treatment makes the inner and outer skin harder than the core.

The Mark IV has only 22 mm of armor on the sides, but this is reinforced by an additional thickness of 22 mm covering the whole fighting and driving compartments. These additional plates are of the machinable type, and the hardness of this plate was found to be 370 Brinell. The bolts holding this extra armor in place are weak, and it was found that the threads stripped easily.

The table below shows the ranges at which the different types of German tank armor are penetrated by standard U.S. and British weapons. The angles of impact are determined by the normal slope of the armor on the tank.

VULNERABILITY OF GERMAN ARMOR PLATE						
	RANGES IN YARDS					
	British 2-pdr		**British**	**U.S.**	**U.S. 75-mm**	
	Standard	**H.V.**	**6-pdr**	**37-mm**	**SAP**	**APC**
Mk. III and IV: 30-mm (old type)						
Lower front plate and turret can be penetrated at	1,300	1,500	Over 2,000	1,600	Over 2,000	
Vizor plate can be penetrated at	1,400	1,600	Over 2,000	1,800	Over 2,000	
Sides can be penetrated at	1,500	1,700	Over 2,000	2,000	Over 2,000	
Mk. IV: 44-mm (reinforced plates)						
Sides can be penetrated at	1,000	1,200	2,000	1,100	Over 2,000	
Mk. III and IV: 62-mm (reinforced plates)						
Lower front plate can be penetrated at	No penetration		500	200	400	1,000
Vizor plate can be penetrated at	No penetration		600	300	500	1,000
Mk. III: 50-mm (new type)						
Lower front plate and turret front can be penetrated at	200	400	800	500	600	1,500
Vizor plate can be penetrated at	200	400	900	600	700	1,700
Sides can be penetrated at	1,500	1,800	Over 2,000	2,000	Over 2,000	

CHAPTER 3:
DEVELOPMENT HISTORY

THE ORIGINS OF THE PANZER IV

The Panzer IV was the brainchild of German general and innovative armored warfare theorist General Heinz Guderian. In concept, it was intended to be a support tank firing mainly high explosive for use against enemy anti-tank guns and fortifications. Ideally, the tank battalions of a panzer division were each to have three medium companies of Panzer IIIs and one heavy company of Panzer IVs.

On 11 January 1934, the German army wrote the specifications for a "medium tractor", and issued them to a number of defense companies. To support the Panzer III, which would be armed with a 37-millimetre (1.46 in) anti-tank gun, the new vehicle would have a short-barrelled 75-millimetre (2.95 in) howitzer as its main gun, and was allotted a weight limit of 24 tonnes (26.46 short

A Panzer Ausf.A rolls into the Sudetenland 1938.

Panzer IV Ausf. C

tons). Development was carried out under the name Begleitwagen ("accompanying vehicle") or BW, to disguise its actual purpose, given that Germany was still theoretically bound by the Treaty of Versailles. MAN, Krupp, and Rheinmetall-Borsig each developed prototypes with Krupp's being selected for further development.

The chassis had originally been designed with a six-wheeled interleaved suspension, but the German Army amended this to a torsion bar system. Permitting greater vertical deflection of the roadwheels, this was intended to improve performance and crew comfort both on- and off-road. However, due to the urgent requirement for the new tank, neither proposal was adopted, and Krupp instead equipped it with a simple leaf spring double-bogie suspension.

The prototype required a crew of five men; the hull contained the engine bay to the rear, with the driver and radio operator, who doubled as the hull machine gunner, seated at the front-left and front-right, respectively. In the turret, the tank commander sat beneath his roof hatch, while the gunner was situated to the left of the gun breech and the loader to the right. The turret was offset 66.5 mm (2.62 in)

Grenadiers crowd aboard a Panzer IV Ausf.J, Russia 1944.

to the left of the chassis center line, while the engine was moved 152.4 mm (6.00 in) to the right. This allowed the torque shaft to clear the rotary base junction, which provided electrical power to turn the turret, while connecting to the transmission box mounted in the hull between the driver and radio operator. Due to the asymmetric layout, the right side of the tank contained the bulk of its stowage volume, which was taken up by ready-use ammunition lockers.

Accepted into service as the Versuchskraftfahrzeug 622 (Vs.Kfz. 622), production began in 1936 at Krupp-Grusonwerke AG's factory at Magdeburg.

AUSF. A TO AUSF. F1

The first mass-produced version of the Panzer IV was the Ausführung A (abbreviated to Ausf. A, meaning "Variant A"), in 1936. It was powered by Maybach's HL 108TR, producing 250 PS (183.87 kW), and used the SGR 75 transmission with five forward gears and one reverse, achieving a maximum road speed of 31 kilometres per hour (19.26 mph). As main armament, the vehicle mounted the Kampfwagenkanone 37 L/24 (KwK 37 L/24) 75 mm (2.95 in) tank gun, which was a low-velocity gun designed to mainly fire high-explosive shells. Against armored targets, firing the Panzergranate (armor-piercing shell) at 430 metres per second (1,410 ft/s) the KwK 37 could penetrate 43 millimetres (1.69 in), inclined at 30 degrees, at ranges of up to 700 metres (2,300 ft). A 7.92 mm (0.31 in) MG 34 machine gun was mounted coaxially with the main gun in the turret, while a second machine gun of the same type was mounted in the front plate of the hull. The Ausf. A was protected by 14.5 mm (0.57 in) of steel armor on the front plate of the chassis, and 20 mm (0.79 in) on the turret. This was capable only of stopping artillery fragments, small-arms fire, and light anti-tank projectiles.

After manufacturing 35 tanks of the A version, in 1937 production moved to the Ausf. B. Improvements included the replacement of the original engine with the more powerful 300 PS (220.65 kW)

The 300 horsepower Maybach HL 120TRM engine used in most Panzer IV production models.

Maybach HL 120TR, and the transmission with the new SSG 75 transmission, with six forward gears and one reverse gear. Despite a weight increase to 16 t (18 short tons), this improved the tank's speed to 39 kilometres per hour (24 mph). The glacis plate was augmented to a maximum thickness of 30 millimetres (1.18 in), and the hull-mounted machine gun was replaced by a covered pistol port.

Forty-two Panzer IV Ausf. Bs were manufactured before the introduction of the Ausf. C in 1938. This saw the turret armor increased to 30 mm (1.18 in), which brought the tank's weight to 18.14 t (20.00 short tons).[After assembling 40 Ausf. Cs, starting with chassis number 80341 the engine was replaced with the improved HL 120TRM. The last of the 140 Ausf. Cs was produced in August 1939, and production changed to the Ausf. D; this variant,

The short-barreled Panzer IV Ausf. F1.

A good study of the main armament of the Panzer IV Ausf .F taken in 1942 in the Army Group Centre sector.

Panzer IV en-route to the front 1941.

A Panzer IV assisted by Russian prisoners attempts to navigate difficult terrain in the Army Group North sector during 1942.

of which 248 vehicles were produced, reintroduced the hull machine gun and changed the turret's internal gun mantlet to an external one. Again protection was upgraded, this time by increasing side armor to 20 mm (0.79 in). As the German invasion of Poland in September 1939 came to an end, it was decided to scale up production of the Panzer IV, which was adopted for general use on 27 September 1939 as the Sonderkraftfahrzeug 161 (Sd.Kfz. 161).

In response to the difficulty of penetrating British Matilda Infantry tanks during the Battle of France, the Germans had tested a 50 mm (1.97 in) gun—based on the 5 cm PaK 38 L/60 anti-tank gun—on a Panzer IV Ausf. D. However, with the rapid German victory in France, the original order of 80 tanks was canceled before they entered production.

In September 1940 the Ausf. E was introduced. This had 50 millimetres (1.97 in) of armor on the bow plate, while a 30-millimetre (1.18 in) appliqué steel plate was added to the glacis as an interim measure. Finally, the commander's cupola was moved forward into the turret. Older model Panzer IV tanks were retrofitted with these features when returned to the manufacturer for servicing. Two hundred and eighty Ausf. Es were produced between December 1939 and April 1941.

In April 1941 production of the Panzer IV Ausf. F started. It featured 50 mm (1.97 in) single-plate armor on the turret and hull, as opposed to the appliqué armor added to the Ausf. E, and a further increase in side armor to 30 mm (1.18 in). The weight of the vehicle was now 22.3 tonnes (24.6 short tons), which required a corresponding modification of track width from 380 to 400 mm (14.96 to 15.75 in) to reduce ground pressure. The wider tracks also facilitated the fitting of ice sprags, and the rear idler wheel and front sprocket were modified.

The designation Ausf. F was changed in the meantime to Ausf. F1, after the distinct new model, the Ausf. F2, appeared. A total of 464 Ausf. F (later F1) tanks were produced from April 1941 to March 1942, of which 25 were converted to the F2 on the production line.

AUSF. F2 TO AUSF. J

On May 26, 1941, mere weeks before Operation Barbarossa, during a conference with Hitler, it was decided to improve the Panzer IV's main armament. Krupp was awarded the contract to integrate again the same 50 mm (1.97 in) Pak 38 L/60 gun into the turret. The first prototype was to be delivered by November 15, 1941. Within months, the shock of encountering the Soviet T-34 medium and KV-1 heavy tanks necessitated a new, much more powerful tank gun.

In November 1941, the decision to up-gun the Panzer IV to the 50-millimetre (1.97 in) gun was dropped, and instead Krupp was contracted in a joint development to modify Rheinmetall's pending 75 mm (2.95 in) anti-tank gun design, later known as 7.5 cm PaK 40 L/46. Because the recoil length was too long for the tank's turret, the recoil mechanism and chamber were shortened. This resulted in the 75-millimetre (2.95 in) KwK 40 L/43. When firing an armor-piercing shot, the gun's muzzle velocity was increased from 430 m/s (1,410 ft/s) to 990 m/s (3,250 ft/s). Initially, the gun was mounted with a single-chamber, ball-shaped muzzle brake, which provided just under 50% of the recoil system's braking ability. Firing the Panzergranate 39, the KwK 40 L/43 could penetrate 77 mm (3.03 in) of steel armor at a range of 1,830 m (6,000 ft).

The 1942 Panzer IV Ausf. F2 was an upgrade of the Ausf. F, fitted with the KwK 40 L/43 anti-tank gun to counter Soviet T-34 and KV tanks.

The Ausf. F tanks that received the new, longer, KwK 40 L/43 gun were named Ausf. F2 (with the designation Sd.Kfz. 161/1). The tank increased in weight to 23.6 tonnes (26.0 short tons). One hundred and seventy-five Ausf. F2s were produced from March 1942 to July 1942. Three months after beginning production, the Panzer IV. Ausf. F2 was renamed Ausf. G.There was little to no difference between the F2 and early G models.

Panzer IV of various types including Ausf.G and Ausf.F(2) meet and confer in Russia during 1943.

A Panzer IV of the 12.SS-Pz.Division undergoes road repairs in the vicinity of Rouen, summer 1944.

THE CONTEMPORARY VIEW NO. 6
NEW ARMAMENT OF GERMAN Pz.Kw. 4

Extracted from Technical and Tactical Trends No. 20, March IIth, 1943

As previously reported in Tactical and Technical Trends (No. 4, p. 15) recent models of two German tanks, the Pz.Kw. 3 and 4, have been fitted with more powerful armament, as shown in the accompanying sketches. These sketches are based on photographs of German tanks captured by the British in North Africa.

PZ.Kw. IV

The principal armament of this tank is a long-barrelled 75-mm gun, the 7.5-cm Kraftwagenkanone 40 (7.5-cm Kw.K. 40). It is reported that the muzzle velocity is 2,400 feet per second (also reported at 2,620 feet per second), and that 2.44 inches of armor plate can be penetrated at 2,000 yards at an angle of impact of 30 degrees. The long barrel, terminating in a muzzle brake, extends beyond the nose of the tank, and an equilibrator was provided, in the particular tank examined, to balance the consequent muzzle preponderance.

The equilibrator is fixed to the floor of the turret and extends vertically to an attachment near the rear of the piece; it is 6 inches in diameter and 21 1/2 inches long. The gun is also provided with a traveling lock inside the turret. The traveling lock consisted of two steel bars about 1/2 inch by 2 inches and 15 inches in length. There were hardened semi-hemispherical surfaces about 1 1/2 inches in diameter projecting from each end of the steel bars, and these fitted into corresponding indentations on either side of lugs attached to the gun and to the turret roof. The steel bars were connected by two bolts; tightening the bolts provided a very positive lock.

Three types of ammunition were found with this tank: nose-fuzed HE; hollow-charge HE; and armor-piercing HE, this being an armor-piercing shell with a ballistic nose and an HE charge.

The Ausf.G

During its production run from May 1942 to June 1943, the Panzer IV Ausf. G went through further modifications, including another armor upgrade. Given that the tank was reaching its viable limit, to avoid a corresponding weight increase, the appliqué 20-millimetre (0.79 in) steel plates were removed from its side armor, which instead had its base thickness increased to 30 millimetres (1.18 in). The weight saved was transferred to the front, which had a 30-millimetre (1.18 in) face-hardened appliqué steel plate welded (later bolted) to the glacis—in total, frontal armor was now 80 mm (3.15 in) thick. This decision to increase frontal armor was favorably received according to troop reports on November 8, 1942, despite technical problems of the driving system due to added weight. At this point, it was decided that 50% of Panzer IV productions would be fitted with 30 mm thick additional armor plates.

PRODUCTION HISTORY	
Designer	Krupp
Designed	1936
Manufacturer	Krupp, Steyr-Daimler-Puch
Unit cost	103,462 Reichsmarks
Produced	1936–45
Number built	9,200 (estimate)
SPECIFICATIONS (Pz IV Ausf H, 1943)	
Weight	25.0 tonnes (27.6 short tons; 24.6 long tons)
Length	5.92 metres (19 ft 5 in), 7.02 metres (23 ft 0 in) gun forward
Width	2.88 m (9 ft 5 in)
Height	2.68 m (8 ft 10 in)
Crew	5 (commander, gunner, loader, driver, radio operator/ bow machine-gunner)
Armor	10–80 mm (0.39–3.1 in)
Main armament	7.5 cm (2.95 in) KwK 40 L/48 main gun (87 rds.)
Secondary armament	2–3 × 7.92-mm Maschinengewehr 34
Engine	12-cylinder Maybach HL 120 TRM V12, 300 PS (296 hp, 220 kW)
Power/weight	12 PS/t
Transmission	6 forward and 1 reverse ratios
Suspension	Leaf spring
Fuel capacity	470 l (120 US gal)
Operational range	200 km (120 mi)
Speed	42 km/h (26 mph) road, 16 km/h (9.9 mph) off road

THE CONTEMPORARY VIEW NO. 7
ARMOR ARRANGEMENT ON GERMAN TANKS
Extracted from Technical and Tactical Trends No. 29, July 15th 1943

The accompanying sketches show the armor arrangement on current models of the PzKw 2, 3, 4, and 6. These sketches are believed to be accurate and up-to-date. Armor thicknesses (circled figures) are given in millimeters; their equivalent in inches may be found in the article beginning on page 30.

A question mark following some of these figures indicates that definite information is not available. Where two small figures appear in parentheses, it indicates that there are 2 plates at this point; in only 2 instances, namely on the PzKw 3, are the 2 plates separated to form so-called spaced armor.

The armament of these tanks is also shown.

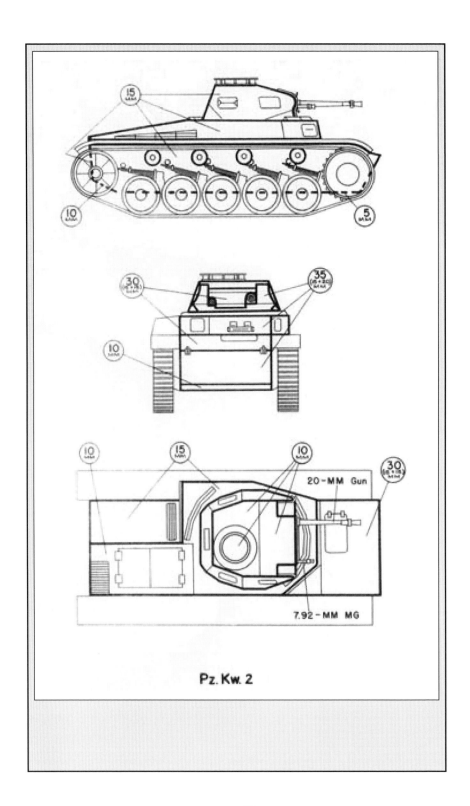

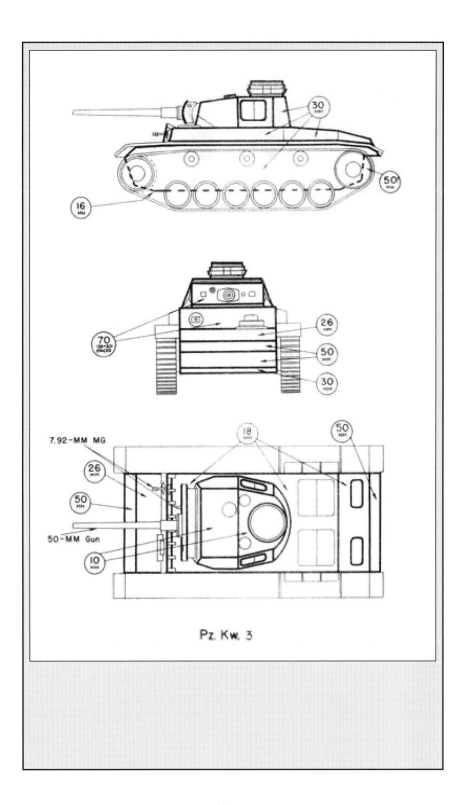

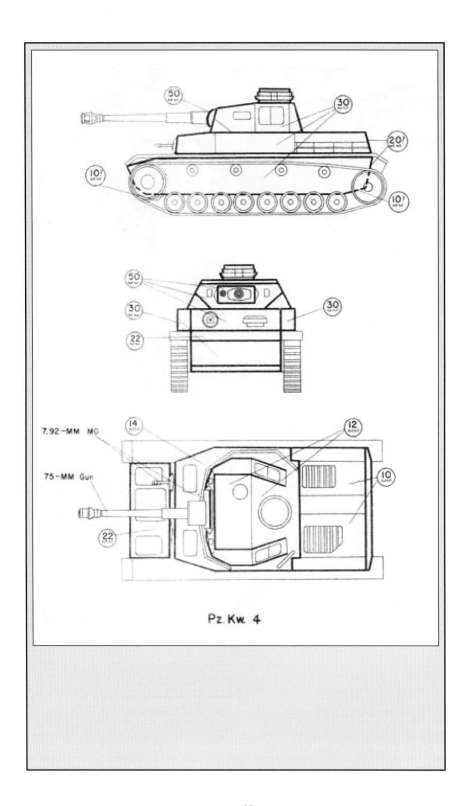

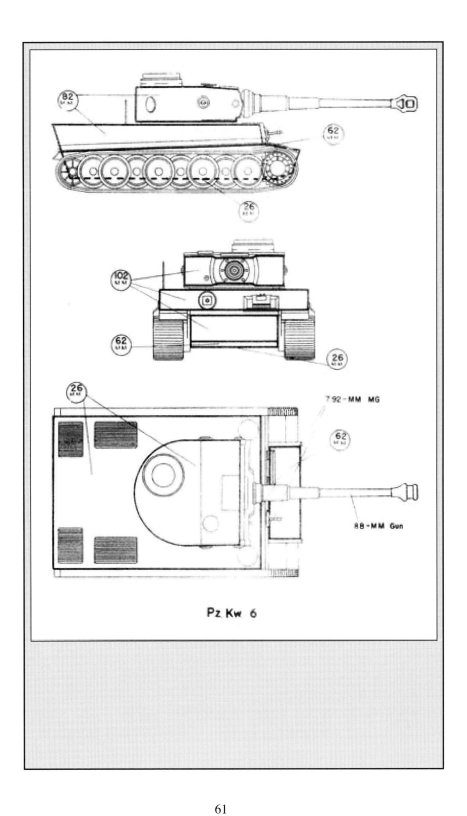

An excellent study of Panzer IV and Panzer II tanks negotiating the Russian terrain during June 1942

Subsequently on January 5, 1943, Hitler decided to make all Panzer IV with 80 mm frontal armor. To simplify production, the vision ports on either side of the turret and on the right turret front were removed, while a rack for two spare road wheels was installed on the track guard on the left side of the hull. Complementing this, brackets for seven spare track links were added to the glacis plate. For operation in high temperatures, the engine's ventilation was improved by creating slits over the engine deck to the rear of the chassis, and cold weather performance was boosted by adding a device to heat the engine's coolant, as well as a starter fluid injector. A new light replaced the original headlight, and the signal port on the turret was removed. On March 19, 1943, the first Panzer IV with Schürzen skirts on its sides and turret was exhibited. The double hatch for the commander's cupola was replaced by a single round hatch from very late model Ausf. G. and the cupola was up-armored as well. In April 1943, the KwK 40 L/43 was replaced by the longer 75-millimetre (2.95 in) KwK 40 L/48 gun, with a redesigned multi-baffle muzzle brake with improved recoil efficiency.

A U.S. report on the German practice of mounting armor skirts (Schürzen) on panzers in WWII, from Tactical and Technical Trends, No. 40, December 16, 1943 is reprinted opposite:

THE CONTEMPORARY VIEW NO. 8
ARMOR SKIRTING ON GERMAN TANKS
Extracted from Technical and Tactical Trends No. 40, December 16th 1943

Panzer IV Ausf.H in the Army Group South sector August 1943.

From both Allied and German sources, reports have come in of additional armored skirting applied to the sides of German tanks and self-moving guns to protect the tracks, bogies and turret. Photographs show such plating on the PzKw 3 and 4, where the plates are hung from a bar resembling a hand-rail running above the upper track guard and from rather light brackets extending outward about 18 inches from the turret.

What appeared to be a 75-mm self-moving gun was

partially protected by similar side plates over the bogies. This armor is reported to be light -- 4 to 6 millimeters (.16 to .24 in) -- and is said to give protection against hollow-charge shells, 7.92-mm tungsten carbide core AT ammunition, and 20-mm tungsten carbide core ammunition. This armor might cause a high-velocity AP shot or shell to deflect and strike the main armor sideways or at an angle, but covering the bogies or Christie wheels would make the identification of a tank more difficult, except at short ranges.

A further U.S. military report on the German use of armor-skirting on tanks was published in Tactical and Technical Trends, No. 42, January 13, 1944.

THE CONTEMPORARY VIEW NO. 9
ENEMY USE OF SKIRTING ON TANKS
Extracted from
Technical and Tactical Trends
No. 42, January 13th 1944

An examination of German Pz Kw 3 and 4 tanks in Sicily, and a number of SP guns has confirmed prior reports that the Germans are using skirting both around the turret and along the sides of the hull. A prior reference to enemy use of armor skirting on German tanks may be found in Tactical and Technical Trends No. 40, p. 11.

On one Pz Kw tank, 1/4-inch mild steel plates were placed around the sides and rear of the turret, and extended from the turret top to the bottom, almost flush with the top of the superstructure. The front edges on both sides had been turned in, so as to line up with the front of the turret, thus filling the space between the turret and the outer mild steel plate. Doors are provided in the outer plate immediately opposite the doors of the turret. The plate is bolted on to brackets by 3/8-inch bolts and studs. The plates stand out about 18 inches from the top and 12 inches from the bottom of the turret. The depth of the plate is approximately 20 inches.

The skirting of 3/16 inch mild steel plates is in sections of 3 feet 9 inches x 3 feet 3 inches. It extends from the top of the superstructure to about the tops of the bogies, and for the full length of the hull. The sections are held in place by slots in them which match the supporting clips on a 1/4-inch angle-iron rail, welded on to the top of the superstructure and extending the full length of the hull, and by 5 brackets bolted on to the track mudguards. The angle-iron is spaced about 15 inches outwards away from the hull, and the brackets about 8 inches away from the mudguards.

Three other Pz Kw 4 tanks, similarly equipped with skirting were also seen, and a Pz Kw 3 tank had both sides completely covered with sheets of 3/16 inch boiler plate extending the whole length of the tank, and reaching from turret-top level to the tops of the bogies.

The 7.5-cm Stu.K. 42 SP equipment on a Pz Kw 3 chassis has been seen with similar additional side

plates. The plates, which extend vertically from the top of the equipment to the tops of the bogies, and laterally from the fifth bogie to the rear of the front-drive sprocket, are in three sections, the front section being cut to conform roughly with the shape of the equipment. A 15-cm s.F.H. 18 on Pz Kw 4 tank chassis is also reported to have been similarly equipped.

It would appear from available information that the use of spaced skirting on German armored vehicles and self-propelled guns is being adopted as standard practice. The fact that the side plates are in sections and held in place by clips suggests that they are detachable. This would, of course, be a great convenience in loading for transportation by rail.

It is believed that the skirting is designed to cause premature explosion of hollow charge, HE and AP HE shell, and thus minimize their effect. Although the plates have been described as mild steel, other sources have erroneously described them as armor.

> Particular attention is drawn to the difficulty of recognition of tanks and SP equipments with this extensive skirting. Almost all of the features which are of primary importance in identification are obscured (see last sentence, Tactical and Technical Trends, No. 40, p. 11).

THE AUSF. H

The next version, the Ausf. H, began production in April 1943 and received the designation Sd. Kfz. 161/2. This variant saw the integrity of the glacis armor improved by manufacturing it as a single 80-millimetre (3.15 in) plate. To prevent adhesion of magnetic anti-tank mines, which the Germans feared would be used in large numbers by the Allies, Zimmerit paste was added to all the vertical surfaces of the tank's armor.

The vehicle's side and turret were further protected by the addition of 5-millimetre (0.20 in) side-skirts and 8-millimetre (0.31 in) turret skirts. During the Ausf. H's production run its rubber-tired return rollers were replaced with cast steel; the hull was fitted with triangular supports for the easily-damaged side-skirts. A hole in the roof, designed for the Nahverteidigungswaffe, was plugged by a circular armored plate due to shortages of this weapon. These modifications meant that the tank's weight jumped to 25 tonnes (27.56 short tons), reducing its speed, a situation not improved by the decision to adopt the Panzer III's six-speed SSG 77 transmission, which was inferior to that of earlier-model Panzer IVs.

The Ausf. J was the final production model, and was greatly simplified compared to earlier variants to speed construction. This shows an exported Finnish model.

Despite addressing the mobility problems introduced by the previous model, the final production version of the Panzer IV—the Ausf. J—was considered a retrograde from the Ausf. H. Born of German necessity to replace heavy losses, it was greatly simplified

Panzer IV Auf.J with missing sideskirts Russia 1944.

to speed production.

The electric generator that powered the tank's turret traverse was removed, so the turret had to be rotated manually. The space was later used for the installation of an auxiliary 200-litre (44 imp gal) fuel tank; road range was thereby increased to 320 kilometres (198.84 mi), The pistol and vision ports in the turret were removed, and the engine's radiator housing was simplified by changing the slanted sides to straight sides. In addition, the cylindrical muffler was replaced by two flame-suppressing mufflers. By late 1944, Zimmerit was no longer being applied to German armored vehicles, and the Panzer IV's side-skirts had been replaced by wire mesh, while to further speed production the number of return rollers was reduced from four to three.

In a bid to augment the Panzer IV's firepower, an attempt was made to mate a Panther turret—carrying the longer 75 mm (2.95 in) L/70 tank gun—to a Panzer IV hull. This was unsuccessful, and confirmed that the chassis had, by this time, reached the limits of its adaptability in both weight and available volume.

THE CONTEMPORARY VIEW NO. 10
GERMAN HOLLOW-CHARGE AMMUNITION FOR 75-MM TANK GUN

Extracted from
Technical and Tactical Trends
No. 19, February 25th, 1943

A sketch showing the details of the hollow-charge round for the German 7.5-cm KwK (75-mm tank gun) accompanies this report. The German nomenclature for this ammunition is 7.5-cm Pz. Gr. Patr. 38 KwK.

The round is of the fixed type. The cartridge case and the weight and type of propellant are similar to those for the other types of 75-mm antitank gun ammunition. The shell is fitted with a threaded hemispherical cap into which is screwed a small nose percussion fuze. From the nose fuze, a central tube runs down to a booster which is situated in the base of the shell. This booster consists of a detonator set in penthrite wax, the whole being contained in a perforated container. The bursting charge consists of three blocks of Hexagen (Trimethylene Trinitramine) the front one of which is concave, as shown in the sketch. The blocks are contained in waxed paper and are cemented into the shell.

The operation of the Aufschlag Zunder or percussion fuze (A.Z. 38-type fuze) is simple. The striker is held

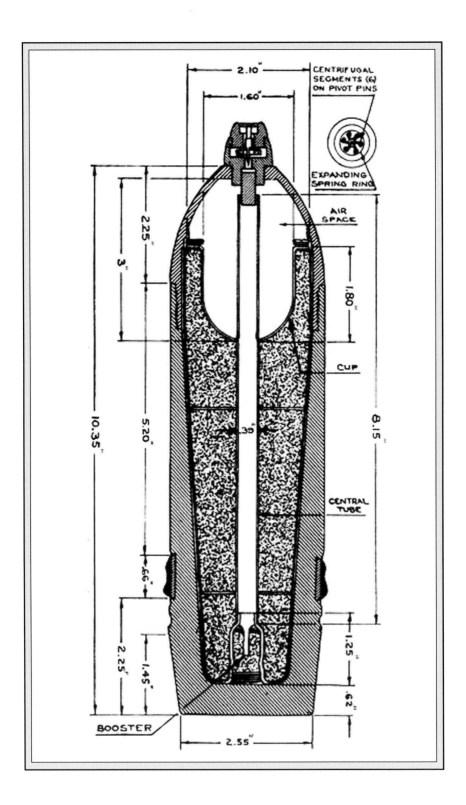

off the detonator assembly by six centrifugal segments which are surrounded by an expanding spring ring. After the shell has left the gun, centrifugal force causes the clock spring and the safety blocks to open, thus freeing the striker. Upon impact, the striker is driven onto the detonator. The detonation passes down the central tube to initiate the booster. This in turn initiates the bursting charge.

The shell is painted white and has black markings. The weight of the shell is 4.5 kilograms, and that of the bursting charge 450 grams.

Comments: This is another instance of the use of hollow-charge ammunition to increase the armor-shattering effect of a gun of comparatively low muzzle velocity. No data is available at this time concerning the performance of this type of projectile against armor at various ranges.

Panzer IV Ausf.D advancing towards the front October 1941

PANZER IV PRODUCTION BY YEAR		
Date	Number of Vehicles	Variant (Ausführung or Ausf.)
1937–1939	262	A – D
1940	278-386	E
1941	467-769	E, F1, F2, G
1942	est. 880	G
1943	3,013	G, H
1944	3,125	J
1945	est. 435	J
Total	9,870	A - J

The Panzer IV was originally intended to be used only on a limited scale, so initially Krupp was its sole manufacturer. Prior to the Polish campaign, only 262 Panzer IVs were produced: 35 Ausf. A; 42 Ausf. B; 140 Ausf. C; and 45 Ausf. D. After the invasion of Poland, and with the decision to adopt the tank as the mainstay of Germany's armored divisions, production was extended to the Nibelungenwerke factory (managed by Steyr-Daimler-Puch) in the Austrian city of St. Valentin. Production increased as the Ausf. E was introduced, with 223 tanks delivered to the German army. By 1941, 462 Panzer IV Ausf. Fs had been assembled, and the up-gunned Ausf. F2 was entering production. The yearly production total had more than quadrupled since the start of the war.

As the later Panzer IV models emerged, a third factory, Vomag (located in the city of Plauen), began assembly. In 1941 an average of 39 tanks per month were built, and this rose to 83 in 1942, 252 in 1943, and 300 in 1944. However, in December 1943, Krupp's factory was diverted to manufacture the Sturmgeschütz IV, and in the spring of 1944 the Vomag factory began production of the Jagdpanzer IV, leaving the Nibelungenwerke as the only plant still assembling the Panzer IV. With the slow collapse of German industry under pressure from Allied air and ground offensives - in October 1944 the Nibelungenwerke factory was severely damaged during a bombing raid — by March and April 1945 production had fallen to pre-1942 levels, with only around 55 tanks per month coming off the assembly lines.

THE EXPORT OF THE PZ IV

The Panzer IV was the most exported German tank of the Second World War. In 1942 Germany delivered 11 tanks to Romania and 32 to Hungary, many of which were lost on the Eastern Front between the final months of 1942 and the beginning of 1943. Romania received approximately 120 Panzer IV tanks of different models throughout the entire war. To arm Bulgaria, Germany supplied 46 or 91 Panzer IVs, and offered Italy 12 tanks to form the nucleus of a new armored division. These were used to train Italian crews while Italian dictator Benito Mussolini was deposed, but were retaken by Germany during its occupation of Italy in mid-1943. The Spanish government petitioned for 100 Panzer IVs in March 1943, but only 20 were ever delivered, by December. Finland bought 30, but received only 15 Panzer IVs in 1944, and the same year a second batch of 62 or 72 were sent to Hungary (although 20 of these were diverted to replace German losses). In total some 297 Panzer IVs of all models were delivered to Germany's allies.

A PzKpfw IV Ausf. H of the 12th Panzer Division operating on the Eastern Front in the USSR, 1944.

Panzer IV Ausf.H on manouvers in Northern France during the summer of 1943.

COMBAT HISTORY

The Panzer IV was the only German tank to remain in both production and combat throughout World War II, and measured over the entire war it comprised 30% of the Wehrmacht's total tank strength. Although in service by early 1939, in time for the occupation of Czechoslovakia, at the start of the war the majority of German armor was made up of obsolete Panzer Is and Panzer IIs. The Panzer I in particular had already proved inferior to Soviet tanks, such as the T-26, during the Spanish Civil War.

WESTERN FRONT AND NORTH AFRICA (1939–1942)

When Germany invaded Poland on 1 September 1939, its armored corps was composed of 1,445 Panzer Is, 1,223 Panzer IIs, 98 Panzer IIIs and 211 Panzer IVs; the more modern vehicles amounted to less than 10% of Germany's armored strength. The 1st Panzer Division had a roughly equal balance of types, with 17 Panzer Is, 18 Panzer

IIs, 28 Panzer IIIs, and 14 Panzer IVs per battalion. The remaining panzer divisions were heavy with obsolete models, equipped as they were with 34 Panzer Is, 33 Panzer IIs, 5 Panzer IIIs, and 6 Panzer IVs per battalion. Although the Polish army possessed less than 200 tanks capable of penetrating the German light tanks, Polish anti-tank guns proved more of a threat, reinforcing German faith in the value of the close-support Panzer IV.

Despite increasing production of the medium Panzer IIIs and IVs prior to the German invasion of France on 10 May 1940, the majority of German tanks were still light types. According to Heinz Guderian, the Wehrmacht invaded France with 523 Panzer Is, 955 Panzer IIs, 349 Panzer IIIs, 278 Panzer IVs, 106 Panzer 35(t)s and 228 Panzer 38(t)s. Through the use of tactical radios and superior tactics, the Germans were able to outmaneuver and defeat French and British armor. However, Panzer IVs armed with the KwK 37 L/24 75-millimetre (2.95 in) tank gun found it difficult to engage French tanks such as Somua S35 and Char B1. The Somua S35 had a maximum armor thickness of 55 mm (2.17 in), while the

A detachment of Panzer IV Ausf.G being transported by rail 1942.

KwK 37 L/24 could only penetrate 43 mm (1.69 in) at a range of 700 m (2,296.59 ft). Likewise, the British Matilda Mk II was heavily armored, with at least 70 mm (2.76 in) of steel on the front and turret, and a minimum of 65 mm on the sides.

Although the Panzer IV was deployed to North Africa with the German Afrika Korps, until the longer gun variant began production, the tank was outperformed by the Panzer III with respect to armor penetration. Both the Panzer III and IV had difficulty in penetrating the British Matilda II's thick armor, while the Matilda's 40-mm QF 2 pounder gun could knock out either German tank; its major disadvantage was its low speed. By August 1942, Rommel had only received 27 Panzer IV Ausf. F2s, armed with the L/43 gun, which he deployed to spearhead his armored offensives. The longer gun could penetrate all American and British tanks in theater at ranges of up to 1,500 m (4,900 ft). Although more of these tanks arrived in North Africa between August and October 1942, their numbers were insignificant compared to the amount of matériel shipped to British forces.

The Panzer IV also took part in the invasion of Yugoslavia and the invasion of Greece in early 1941.

EASTERN FRONT (1941–1945)

With the launching of Operation Barbarossa on 22 June 1941, the unanticipated appearance of the KV-1 and T-34 tanks prompted an upgrade of the Panzer IV's 75 mm (2.95 in) gun to a longer, high-velocity 75 mm (2.95 in) gun suitable for antitank use. This meant that it could now penetrate the T-34 at ranges of up to 1,200 m (3,900 ft) at any angle. The 75 mm (2.95 in) KwK 40 L/43 gun on the Panzer IV could penetrate a T-34 at a variety of impact angles beyond 1,000 m (3,300 ft) range and up to 1,600 m (5,200 ft). Shipment of the first model to mount the new gun, the Ausf. F2, began in spring 1942, and by the summer offensive there were around 135 Panzer IVs with the L/43 tank gun available. At the time, these were the only German tanks that could defeat the Soviet T-34 or KV-1. They

A Panzer IV at rest next to a Panje hut on the Russian steppe, September 1942.

12. A Panzer IV Ausf.D fords a Russian river, summer 1941

Panzer IV Ausf.H of Army Group Centre, Poland 1944.

played a crucial role in the events that unfolded between June 1942 and March 1943, and the Panzer IV became the mainstay of the German panzer divisions. Although in service by late September 1942, the Tiger I was not yet numerous enough to make an impact and suffered from serious teething problems, while the Panther was not delivered to German units in the Soviet Union until May 1943. The extent of German reliance on the Panzer IV during this period is reflected by their losses; 502 were destroyed on the Eastern Front in 1942.

The Panzer IV continued to play an important role during operations in 1943, including at the Battle of Kursk. Newer types such as the Panther were still experiencing crippling reliability problems that restricted their combat efficiency, so much of the effort fell to the 841 Panzer IVs that took part in the battle. Throughout 1943, the German army lost 2,352 Panzer IVs on the Eastern Front; some divisions were reduced to 12–18 tanks by the end of the year. In 1944, a further 2,643 Panzer IVs were destroyed, and such losses were becoming increasingly difficult to replace. By the last year of the war, the Panzer IV was outclassed by the upgraded T-34-85, which had an 85 mm (3.35 in) gun, and other late-model Soviet tanks such as the 122 mm (4.80 in)-armed IS-2 heavy tank. Nevertheless, due to a shortage of replacement Panther tanks, the Panzer IV continued to form the core of Germany's armored divisions, including elite units such as the II SS Panzer Corps, through 1944.

In January 1945, 287 Panzer IVs were lost on the Eastern Front. It is estimated that combat against Soviet forces accounted for 6,153 Panzer IVs, or about 75% of all Panzer IV losses during the war.

WESTERN FRONT (1944–1945)

Panzer IVs comprised around half of the available German tank strength on the Western Front prior to the Allied invasion of

British officers inspect a German Pzkw-IV knocked out in France in June 1944 by the Durham Light Infantry.

A dug-in Panzer IV of the 22nd Panzer Regiment, photographed near Lebisey after being knocked out during Operation Charnwood.

A Panzer IV wreck lies abandoned, Normandy August 1944.

Normandy on June 6, 1944. Most of the 11 panzer divisions that saw action in Normandy initially contained an armored regiment of one battalion of Panzer IVs and another of Panthers, for a total of around 160 tanks, although Waffen-SS panzer divisions were generally larger and better-equipped than their Heer counterparts. Regular upgrades to the Panzer IV had helped to maintain its reputation as a formidable opponent. Despite overwhelming Allied air superiority, the Norman bocage countryside in the US sector heavily favored defense, and German tanks and anti-tank guns inflicted horrendous casualties on Allied armor during the Normandy campaign. On the offensive, however, the Panzer IVs, Panthers and other armored vehicles proved equally vulnerable in the bocage, and counter-attacks rapidly stalled in the face of infantry-held anti-tank weapons, tank destroyers and anti-tank guns, as well as the ubiquitous fighter bomber aircraft. That the terrain was highly unsuitable for tanks was illustrated by the constant damage suffered to the side-skirts of the Ausf. H's; essential for defence against shaped charge anti-tank weapons such as the British PIAT, all German armored units were "exasperated" by the way these were torn off during movement through the dense orchards and hedgerows.

The Allies had also been developing lethality improvement programs of their own; the widely-used American-designed M4 Sherman medium tank, while mechanically reliable, suffered from thin armor and an inadequate gun. Against earlier-model Panzer IVs, it could hold its own, but with its 75 mm M3 gun, struggled against the late-model Panzer IV (and was unable to penetrate the frontal armor of Panther and Tiger tanks at virtually any range). The late-model Panzer IV's 80 mm (3.15 in) frontal hull armor could easily withstand hits from the 75 mm (2.95 in) weapon on the Sherman at normal combat ranges, though the turret remained vulnerable.

The British up-gunned the Sherman with their highly effective QF 17 pounder anti-tank gun, resulting in the Firefly; although this was the only Allied tank capable of dealing with all current German tanks at normal combat ranges, few (about 300) were available in time for the Normandy invasion. The other British tank with the 17 pdr gun could not participate in the landings and had to wait for port facilities. It was not until July 1944 that American Shermans, fitted with the 76-mm (3-inch) M1 tank gun, began to achieve a parity in firepower with the Panzer IV.

However, despite the general superiority of its armored vehicles, by August 29, 1944, as the last surviving German troops of Fifth Panzer Army and Seventh Army began retreating towards Germany, the twin cataclysms of the Falaise Pocket and the Seine crossing had cost the Wehrmacht dearly. Of the 2,300 tanks and assault guns it had committed to Normandy (including around 750 Panzer IVs), over 2,200 had been lost. Field Marshal Walter Model reported to Hitler that his panzer divisions had remaining, on average, five or six tanks each.

During the winter of 1944–45, the Panzer IV was one of the most widely used tanks in the Ardennes offensive, where further heavy losses—as often due to fuel shortages as to enemy action—impaired major German armored operations in the West thereafter. The Panzer IVs that took part were survivors of the battles in France between June and September 1944, with around 260 additional Panzer IV Ausf. Js issued as reinforcements.

OTHER USERS

In the 1960s Syria received a number of Panzer IVs from the French, replacing the turret's machine gun with a Soviet-made 12.7-millimetre (0.50 in) machine gun. These were used to shell Israeli settlements below the Golan Heights, and were fired upon during the 1965 "Water War" by Israeli Centurion tanks. Syria received 17 more Panzer IVs from Spain, which saw combat during the Six-Day War in 1967.

The Finns bought 15 new Panzer IV Ausf J in 1944, for 5,000,000 Finnish markkas each (about twice the production price). The tanks arrived too late to see action against the Soviets, but were instead used against the Germans in the Lapland War. After the war, they

A Panzer IV of the SS Leibstandarte AH rolls through Italy during 1943.

A Syrian Panzer IV Ausf. G, captured during the Six-Day War, on display in the Yad La-Shiryon Museum, Israel.

served as training tanks, and one portrayed a Soviet KV-1 tank in the movie The Unknown Soldier in 1955.

After 1945, Bulgaria incorporated its surviving Panzer IVs in defensive bunkers as gunpoints on the border with Turkey, along with T-34 turrets. This defensive line known as the "Krali Marko Line", remained in use until the fall of communism in 1989.

Most of the tanks Romania had received were lost in 1944 and 1945 in combat. These tanks, designated T4 in the army inventory, were used by the 2nd Armoured Regiment. On 9 May 1945 only two Panzer IV were left. Romania received another 50 Panzer IV tanks from the Red Army after the end of the war. These tanks were of different models and were in very poor shape. Many of them were missing parts and the side skirts. The T4 tanks remained in service until 1950, when the Army decided to use only Soviet equipment. By 1954, all German tanks were scrapped.

VARIANTS

In keeping with the wartime German design philosophy of mounting an existing anti-tank gun on a convenient chassis to give mobility, several tank destroyers and infantry support guns were built around the Panzer IV hull. Both the Jagdpanzer IV, initially armed with the 75-millimetre (2.95 in) L/48 tank gun, and the Krupp-manufactured Sturmgeschütz IV, which was the casemate of the Sturmgeschütz III mounted on the body of the Panzer IV, proved highly effective in defense. Cheaper and faster to construct than tanks, but with the disadvantage of a very limited gun traverse, around 1,980 Jagdpanzer IV's and 1,140 Sturmgeschütz IVs were produced. The Jagdpanzer IV eventually received the same 75 millimeter L/70 gun that was mounted on the Panther.

Another variant of the Panzer IV was the Panzerbefehlswagen IV (Pz.Bef.Wg. IV) command tank. This conversion entailed the installation of additional radio sets, mounting racks, transformers, junction boxes, wiring, antennas and an auxiliary electrical generator. To make room for the new equipment, ammunition stowage was reduced from 87 to 72 rounds. The vehicle could coordinate with nearby armor, infantry or even aircraft. Seventeen Panzerbefehlswagen were converted from Ausf. J chassis, while another 88 were based on refurbished chassis.

The Panzerbeobachtungswagen IV (Pz.Beob.Wg. IV) was an artillery observation vehicle built on the Panzer IV chassis. This,

Pzkw-IV in Belgrade Military Museum, Serbia.

The Wirbelwind armored anti-aircraft vehicle.

A Jagdpanzer IV/48 tank destroyer, based on the Panzer IV chassis, mounting the 75 mm PaK L/48 anti-tank gun.

A Sturmpanzer IV Brummbär infantry-support gun (Casemate MG variant (flexible mount)).

too, received new radio equipment and an electrical generator, installed in the left rear corner of the fighting compartment. Panzerbeobachtungswagens worked in cooperation with Wespe and Hummel self-propelled artillery batteries.

Also based on the Panzer IV chassis was the Sturmpanzer IV Brummbär 150-millimetre (5.91 in) infantry-support self-propelled gun. These vehicles were primarily issued to four Sturmpanzer units (Numbers 216, 217, 218 and 219) and used during the battle of Kursk and in Italy in 1943. Two separate versions of the Sturmpanzer IV existed, one without a machine gun in the mantlet and one with a machine gun mounted on the mantlet of the casemate. Furthermore, a 105-millimetre (4.13 in) artillery gun was mounted in an experimental turret on a Panzer IV chassis. This variant was called the Heuschrecke, or Grasshopper. Another 105 mm artillery/anti-tank prototype was the 10.5 cm K (gp.Sfl.) nicknamed Dicker Max.

Four different self-propelled anti-aircraft vehicles were built on the Panzer IV hull. The Flakpanzer IV Möbelwagen was armed with a 37-millimetre (1.46 in) anti-aircraft cannon; 240 were built between 1944 and 1945. In late 1944 a new Flakpanzer, the Wirbelwind, was designed, with enough armor to protect the gun's crew and a rotating turret, armed with the 20mm quadmount Flakvierling anti-aircraft cannon system; at least 100 were manufactured. Sixty-five similar vehicles were built, named the Ostwind, but with a single 37-millimetre (1.46 in) anti-aircraft cannon instead. This vehicle was designed to replace the Wirbelwind. The final model was the Flakpanzer IV Kugelblitz, of which only five were built. This vehicle featured a covered turret armed with twin 30-millimetre (1.18 in) anti-aircraft cannons.

Although not a direct modification of the Panzer IV, some of its components, in conjunction with parts from the Panzer III, were utilized to make one of the most widely-used self-propelled artillery chassis of the war—the Geschützwagen III/IV. This chassis was the basis of the Hummel artillery piece, of which 666 were built, and also the 88 millimetres (3.46 in) gun armed Nashorn tank destroyer, with 473 manufactured. To resupply self-propelled howitzers in the field, 150 ammunition carriers were manufactured on the Geschützwagen III/IV chassis.

Panzer IV of the 4th Panzer Division (Panzerregiment 35)

More from the same series

Most books from the 'Hitler's War Machine' series are edited and endorsed by Emmy Award winning film maker and military historian Bob Carruthers, producer of Discovery Channel's Line of Fire and Weapons of War and BBC's Both Sides of the Line. Long experience and strong editorial control gives the military history enthusiast the ability to buy with confidence.

Tiger I in Combat
Tiger I Crew Manual
Panzers at War 1939-1942
Panzers at War 1943-1945
Wolf Pack - the U boats
Poland 1939
Luftwaffe Combat Reports
Sturmgeschütze
German Artillery in Combat
Panzer Combat Reports
The Panther V in Combat
German Tank Hunters
The Afrika Korps in Combat
Panzers I & II
Panzer III
Panzer IV

For more information visit www.pen-and-sword.co.uk